The Illustrated Guide to

Antique Writing Instruments

**Stuart Schneider
& George Fischler**

Revised & Expanded 3rd Edition

Schiffer Publishing Ltd

4880 Lower Valley Road, Atglen, PA 19310 USA

Revised price guide: 2000
Copyright © 1994, 1997 & 2000 by Stuart Schneider & George Fischler
Library of Congress Catalog Card Number: 99-67572

ISBN: 0-7643-0980-3
Printed in Hong Kong
1 2 3 4

Published by Schiffer Publishing Ltd.
4880 Lower Valley Road
Atglen, PA 19310
Phone: (610) 593-1777; Fax: (610) 593-2002
E-mail: Schifferbk@aol.com
Please visit our web site catalog at
www.SCHIFFERBOOKS.com

In Europe, Schiffer books are distributed by
Bushwood Books
6 Marksbury Avenue Kew Gardens
Surrey TW9 4JF England
Phone: 44 (0)181 392-8585;
Fax: 44 (0)181 392-9876
E-mail: Bushwd@aol.com

This book may be purchased from the publisher.
Include $3.95 for shipping. Please try your bookstore first.
We are interested in hearing from authors with book ideas on related subjects.
You may write for a free printed catalog.

Contents

Introduction

This book is for collectors and dealers alike. It will introduce the beginning pen collector to the hobby of pen collecting and answer questions asked by pen dealers about the values of pens. It will identify pens and provide the opportunity to examine many of the pens that were produced.

There are great pens and no-so-great pens. If a pen appears here, it is either interesting, collectible, or a good example of a particular maker's pen. There are dozens of pens that were not illustrated, not because they were bad pens, but because illustrating every pen made would take a multivolume encyclopedia.

The hobby of collecting fountain pens has grown tremendously and the focus of collecting has changed. The urge to have every pen made has been replaced by collectors seeking out one or two styles or lines of fountain pens. Beginning collectors can no longer amass pens in the quantities that the early collectors could. Pens are becoming harder to find and prices are rising, so collectors must work harder to build their collection. Almost any new collector/user, however, can put together a representative collection in a short time. A few collectors can show collections of 1500 pens, but the average collector has under 25.

Fountain pen makers produced over a million pens a year from the 1910s to the 1940s. The ballpoint pen arrived in 1945 and was eventually proven reliable (the first ones wrote for several weeks and then died). Millions of fountain pens were put in drawers and forgotten.

Four particular pen manufacturers are especially revered. The "big four" companies were the Wahl-Eversharp Pen Company, Sheaffer Pen Company, Parker Pen Company, and Waterman Pen Company.

History

THE L.E. WATERMAN PEN COMPANY
1884-Present

The oldest of the "big four" pen companies was the L.E. Waterman Pen Company of New York. Begun in 1884, using Lewis E. Waterman's patented (February 12, 1884) "Ideal" fountain pen, this company went on to become the largest American pen maker. Waterman started with his patent for an improved feed design. He knew that the simple channel in the "feed" from the ink reservoir to the nib had to serve two masters - the flow of ink out and the flow of air in. A way was devised for the air to enter the pen as the ink flowed out by cutting three razor thin slits in the bottom of the channel leading to the nib. Capillary action brought the ink to the nib's point and the channel now allowed air to enter the reservoir as needed. The pen was a success, and with Waterman's keen use of advertising, the Waterman Pen Company took off, producing some of the most attractive pens ever made, with designs in silver, gold and hard rubber.

Waterman's pen business began in 1883 in a small office at 136 Fulton Street in New York City. Pen parts were ordered from local jobbers and Waterman, who had been an insurance salesman, assembled and sold the pens. In March of 1884, Waterman and his best customer, Asa Shipman, formed a partnership known as the Ideal Pen Company and moved into Shipman's offices at 10 Murray Street. The increased production to about 200 pens that year. Waterman and Shipman, for whatever reason, dissolved the partnership in late 1884 with Waterman remaining at the Murray Street location until he could obtain a new manufacturing site. Early in 1885, Waterman moved to 155 Broadway. To raise capital to pay off debts and expand production Waterman formed the L.E. Waterman Pen Company, a corporation, on November 11, 1887. The Company's business grew at a slow but steady rate and the offices were moved to keep up with the expansion. It took over the offices at 157 Broadway in 1893, then moved to new quarters at 173 Broadway in 1902, and 191 Broadway in 1917.

In 1901, when L.E. Waterman died, at age 64, his company was selling over 1000 pens a day. The company continued to make a variety of attractive pens Waterman's first plastic pens were the Patrician (1929-1938), a large men's pen, and the #94's (1930-1938). These plastic pens used the lever filler system with the exception of the Ink-Vue which used a variation of the lever filler. The Hundred Year pen (1939-1946) designed by John Vassos, followed the Ink-Vue. It was guaranteed for 100 years. Other models made during this period were the Commando (1942-1949) which was basically a 100 Year pen without the guarantee.

Waterman Numbering System

About 1898 Waterman began using a standard numbering system to describe its pens. This system was modified slightly in 1917 and lasted until the 1930s with exceptions creeping in after 1927. These numbers were stamped on the end of the pen as follows:

Hundreds column
2xx Sterling silver barrel covered
3xx Gold barrel covered
4xx Sterling silver barrel & cap
5xx Solid gold barrel & cap
6xx Mother of pearl (after 1917, two gold bands on barrel)
7xx One 14kt gold band on barrel
8xx One 14kt gold band on cap
9xx One 14kt gold band on top of cap

A zero before the hundred's column means gold filled

Tens column
0x Straight cap
1x Cone cap
2x Taper cap
3x Baby size (special short length)
4x Desk pen (after 1917, Safety pen)
5x Lever filler
6x Lever filler w/slip cap
7x Eyedropper w/screw cap
8x Pump filler

Units column
This is the nib size

Additional Markings

 1/2 indicates that the pen is thinner
 V indicates that the pen is short
 SF = Self filler
 S = Safety
 P = Pump
 POC = Pocket
 PSF = Pocket self filler
 X = Larger barrel than standard for that nib size
 VP = Vest pocket

 For example, a #0552 1/2 indicates a thin, gold filled, lever filler pen with a #2 nib. A #404 is a silver, straight cap pen with a #4 nib and a #48 is a lever filler pen with a #8 nib. About 1928, Waterman began making pens that did not use the above system such as #7s, #5s, #94s, #92s, and #32s.

Waterman Filling Systems

 During the Waterman years, different systems were devised to improve the ease and neatness of filling the pen with ink. Some of the systems were extremely short-lived, making these pens hard to find and very desirable to the collector.

 Eyedropper Filler - 1884-1928. The first and simplest system employed by Waterman was the eyedropper filling pens. The "section," composed of the nib, feed and collar, was unscrewed from the barrel and the barrel filled with ink from an eyedropper.

 Pump Filling (1st style) - 1899. The first improvement over the eyedropper filler was the 1899 pump filler. The end of the barrel was removed to expose a rod which was drawn up to pump in the ink. The pen was longer than a standard pen to accommodate the rod.

 Pump Filling (2nd style) - 1903-1926. Filling was accomplished by holding the pen with the nib in the ink bottle, unscrewing the pump mechanism from the end of the barrel and then pumping up and down with short, half inch (1.3cm) strokes. The scarcity of these pens today indicates that it was not popular.

 Safety Pen - 1907-1940s. The safety features were a nib and feed that retracted into the barrel of the pen when the bottom of the pen was turned, and a screw on cap instead of a friction fit cap.

 Sleeve Filler - 1910-1915. In this system a sliding sleeve covered an opening in the barrel. After sliding back the sleeve on the pen barrel the pressure bar was depressed, compressing the sac. Upon release, the sac expanded and ink was drawn in. These pens are uncommon today.

Coin Filler - 1913-1914. The Coin Filler is one of the rarest filling systems. Like the Sleeve Filler, the pen had a rubber sac and pressure bar, but instead of a sleeve, the barrel had a slit into which you pushed a penny of the special Waterman Coin which came with the pen.

Lever Filler - 1915-1955. These had a lever on the side of the barrel. Raising the lever pushed against a metal bar that compressed the ink sac. The lever system eventually replaced all the previous Waterman filling systems and became the fountain pen industry standard.

Cartridge Filler - 1936-Present. Waterman first introduced a pen which filled by means of a glass cartridge in 1936. Starting in the 1950s, plastic cartridges were used.

Ink-Vue Filler - 1935-1940. The Ink-Vue pen was a variation in the lever filler, but was touted as one of the new sacless wonder pens of the 1930s. The sac, depressed by a jointed lever, was used as a means of drawing ink into the sealed barrel.

THE PARKER PEN COMPANY 1888-Present

The Parker Pen company was started by George S. Parker who was born November 1, 1863 in Shullsburg, Wisconsin. He attended telegraphy school in Janesville, Wisconsin in the 1880s and eventually became a teacher there. To supplement his income he sold fountain pens to his students. Parker was also the pen repairman at the school. Like Waterman, he saw problems with the pens that he sold and discovered a way to fix the problem. With the financial backing of W.F. Palmer, an insurance salesman, he started the Parker Pen Company in 1888. His first fountain pen patent was for a form of an over and under feed pen. He applied for the patent and discovered that someone else already held a patent on that idea. Not being discouraged so easily, Parker found the patent owner and bought the rights. Once he owned the rights to the competing patent, his own patent could be registered and was so registered on December 10, 1889. This early Parker pen was manufactured, but pen sales were lackluster.

1892 is acknowledged as the actual beginning of the Parker Pen Company. In that year, Parker designed the "Lucky Curve" pen in answer to another problem with pens. After a pen was left in a drawer or carried in the pocket (before the days of pocket clips, pens were carried in side pockets and kept horizontal) it often dropped a blob of ink onto the paper when it was turned vertically to write. Parker invented a feed that curved against the side of the barrel and channeled ink back into the reservoir. He called this the "Lucky Curve" pen. A patent for the curved feed was issued on January 9, 1894. Production was begun on this over and under feed, Lucky Curve model. Again, sales were less than spectacular. Parker could see that the new Waterman underfed pen was outselling his pen by leaps and bounds. Parker redesigned the Lucky Curve in 1898 as an under feed pen. It received public approval and with the help of advertising, Parker's sales grew steadily.

In late 1921, Parker introduced their *Duofold* pen, a pen guaranteed for 25 years, which evolved from the large "Jack Knife Safety Pen." The first "Big Red" *Duofold* made its appearance in November, 1921 and was a break from the standard, large men's pen. At that time large men's pens were made in a dignified black color, this new *Duofold* was an orange lacquer color. It was an instant success. The first *Duofold* differed from later *Duofolds* in that it did not have a band on the lower portion of the cap. There was a Deluxe model of 1922, available with a wide gold filled band on the cap. The nib on these early pens still said "Lucky Curve" but the barrel imprint said in large letters, *Duofold*. A band was put on the lower portion of the cap in 1923.

Parker began to make pens from plastic in the second half of 1926. Parker's trade name for the plastic was Permanite. With the introduction of the plastic *Duofolds*, new colors made their debut. The *Duofold* line was standardized at 8 colors: Orange, introduced in 1921; Black, introduced in 1922; Green Jade, introduced in 1926; Lapis Blue and Mandarin Yellow, introduced in 1927; Moderne Black & Pearl, introduced in 1928; Modern Green & Pearl, introduced in 1930; and Burgundy Red & Black, introduced in 1930. *Duofold* pens may also be found in Sea Green Pearl Marble, Chocolate Pearl Marble, and Red Pearl Marble. At the end of 1929, Parker streamlined the entire Duofold line, and ceased production of *Duofolds* by 1932.

In 1933, Parker announced a wonderful, new innovation, the "Vacuum Filler" pen, soon to become the famous Parker *Vacumatic* (1933-1948). It had a special rubber ink sac that did not hold ink but acted as the flexible plunger mechanism. The Vacumatic pen was designed by industrial designer, Joseph Platt and was made with a new laminated, striped pearl pattern of rings around the barrel and cap. The clip was also changed to the new Parker "Arrow" clip. Several changed were made in the *Vacumatic* line during its lifetime such as new colors, new sizes, three variations in the vacuum filling system, a modified shape and different clip styles. In 1939, a small "Blue Diamond" was added to the top of the Arrow clip signifying that the pen was guaranteed for life.

Parker's next pen design changed the style and look of all fountain pens to come. In 1941, Parker showed off its new Parker *51*. It had a hooded nib and used a new ink that dried as soon as it touched the paper. In terms of design, it was an incredibly original and fresh idea and ushered in the period where thin pens were the "in" pen. Other pen makers tried to imitate it with almost no success. About 1946, Parker began to turn out over a million *51* pens a year.

SHEAFFER PEN COMPANY 1913-Present

Walter A. Sheaffer was born on July 27, 1867 in Bloomfield, Iowa. He was trained as a jeweler and later worked in his father's jewelry store. After gaining experience there, he opened his own jewelry store in Fort Madison, Iowa in 1906. The store sold and repaired fountain pens. Sheaffer designed the lever filling mechanism in 1907 and patented it in 1908. Several years went by while he sought partners. Sheaffer took the risk and by 1912, started production. The Sheaffer Pen Company was incorporated in Fort Madison, Iowa, by Walter A. Sheaffer (51% owner), Ben Coulson & George Kraker (40% owners), and James Brewster, a banker (9% owner) on January 1, 1913. The lever filler was clean, simple and instantly popular. It became popular almost overnight. Other manufacturers redesigned the lever system and obtained patents that allowed them to produce versions of the lever filler pen.

Sheaffer created another first for the pen industry in 1920 when the *Lifetime* model was introduced. This pen was guaranteed for the life of the original owner and would be repaired or replaced, no mater what type of damage, for free. Identification of the *Lifetime* models was originally made by the *Lifetime* size, a larger pen with a #8 nib. In 1921, Sheaffer began to make pens with the *Lifetime* nib and in 1923, the pens could be identified by the *Lifetime* white dot somewhere on the pen. The large men's *Lifetime* sold for $8.75, a new high price for writing instruments at that time. A Sheaffer *Lifetime* in one's pocket became a status symbol, very much like the large, present-day Mont Blanc pen.

In 1924, Sheaffer again stunned the pen market with the introduction of the first plastic (called "Radite") pen in a stunning Jade Green color. Other companies did not introduce a plastic pen until the second half of 1926 or later. These large green *Lifetime* pens competed admirably with the Parker Big Red and the Waterman Ripple.

Sheaffer streamlined its pen line in 1930, calling the new stream-lined version the *Balance* pen. In 1934, Sheaffer, while still making the lever filler, began making a pump filler in response to the Parker *Vacumatic* which had been introduced in 1933. Sheaffer's pump used a long, thin metal rod attached to several washers to create suction to pull the ink into the barrel. There was no sac in these model. The system may have worked well when the pen was new, but over time, as the rubber washers hardened and the packing deteriorated, the pen would leak or not fill at all. Today they are difficult to repair and are often avoided by collectors.

Another type of pump filler pen was introduced in 1949 - the Touch-down Pen. It had a thick metal tube that slid into the barrel and contained a thin ink sac. When the tube end of the pen was pushed in, the air pressure compressed the sac. Again in response to another pen maker's innovative product - Parker's introduction of the Parker *51* - Sheaffer introduced the "Triumph" nib which was used form 1942 to 1962 alongside its regular nib style. The Triumph was a rather stiff writing nib designed as a small cylinder of gold that rose up into a point.

Another popular innovation, the "Snorkel" pen (1952-1963), introduced in 1952, had a slotted tube that retracted and extended when you twisted the rear of the barrel. The tube went through the center of the feed under the nib. The Snorkel was used on the large "PFM" (Pen For Men) line into the early 1960s.

WAHL-EVERSHARP PEN COMPANY 1914-1957

The Wahl-Eversharp Pen Company started out in an entirely different manner than the preceding pen companies. Wahl was incorporated in September, 1905 as the Wahl Adding Machine Company, making adding machines. It was a corporation that was willing to invest in anything that would show a profit. Wahl entered the writing instrument field by buying the Ever-Sharp Pencil Company in 1914. Ever-Sharp was founded by Tokuji Hayakawa, inventor and producer of a metal mechanical pencil and later founder of the company that became Sharp Electronics. A factory was opened at 1800 Roscoe Street in Chicago, Illinois to make pencils. The productions of these Eversharp pencils was profitable and Wahl looked for a fountain pen company to acquire. After looking at several companies, Wahl acquired the Boston Fountain Pen Company located in Massachusetts in late 1917. Boston Pen was a small, high quality pen company with a collection of valuable patents in addition to its machinery and personnel. The Boston Pen operation was moved to Chicago.

In 1918, Wahl began selling the Wahl Tempoint Pen, which was basically the unchanged Boston pen. In 1921, Wahl introduced an all metal pen to match its line of Eversharp pencils. It maintained its lead in pen sales until the mid 1920s when the other big pen companies began to produce their line of colorful pens. Wahl bought a rubber company and began to produce its own line of colorful pens. In hind sight, this was a terrible mistake, as new lines of plastic pens from Sheaffer and Parker soon made their debut. Wahl was left, not only with a large supply of rubber pen parts, but a large rubber company. It continued to make hard rubber pens and sales began to fall. While not popular when introduced, to the collector these pens made from 1926 to 1929 are some of the best balanced and greatest writing hard rubber pens ever made.

Wahl finally began to produce plastic pens in 1927. In 1928, they introduced the Personal Point pen. In 1929, a small gold seal with two check marks was placed on the cap to indicate that these were their best pens. The "Personal Point" feature of these new pens was that the nib and feed could be unscrewed from the main section of the pen and replaced easily. Wahl produced a simple streamlined version (1930-1931) of their Personal Point pens called the "Equipoised" and followed them with the *Doric* pen (1931-1941). The *Doric's* design was very "Deco" with its tapered 12-sided shape. *Doric* pens used the roller ball clip until 1935 when a simpler Art Deco clip replaced it. The large sized *Dorics* were discontinued in 1936, when Wahl introduced a large metal pen with plastic inserts called the *Coronet* (1936-1941).

Whal Pen Company and Eversharp Pencil Company reorganized themselves into the Eversharp Company in 1940-41 and began production of a new pen that again brought the company around to profitability - The Skyline Pen (1941-1949). The *Skyline* wrote beautifully, looked attractive, and was advertised not to leak in airplanes. The *Skyline* was available in many colors and price ranges, from the simple black to the solid gold "Command Performance" pen.

Towards the end of the company's existence, many new products were introduced - the Symphony (designed by Raymond Lowey), the Envoy, and the Ventura Burp Pen - but none of the was very successful. In December, 1957, Eversharp sold its writing instruments division to Parker Pen.

SWAN PEN: MABIE TODD & CO. 1843-1939

The Mabie Todd Company was the successor to the Mabie, Todd & Bard Company which was the offspring of the Bard Bros. Bard was started in 1843, in New York City, making gold pen points and pen point holders. The company, with new management, was reformed as the Mabie, Todd & Bard Company in 1873. By 1880, they were producing stylographic pens at 17 Maiden Lane, under their own name and were producing pen points for other pen companies. About 1908, the name "Bard" was removed from the company name, although many of the pens still carried that name. Mabie Todd made very high quality eyedropper pens. Most of their pens made before 1920 were manufactured in New York. Mr. Mabie sold his interest in the company to an Englishman named Watts in 1908. The ownership remained mostly American until 1919, when it was sold to a group in Great Britain.

The top of the Mabie Todd line was the Swan Pen and in the late teens, the matching pencil was called the "Fyne Poynt" pencil. The pens of Mabie Todd made with gold and silver overlays, were as much a form of men's jewelry as they were writing instruments. On some Swan pens there is a gold needle-like piece on top of the nib. This was Swan's "Gold Top Feed" (in addition to the bottom fed) which was supposed to guarantee ink flow to the nib, regardless of writing pressure. This feature first appears on some 1880s pens and continued, on some models, into the early 1920s. Lever filler models were started about 1917. Swan's contribution to the "Lifetime" models was the *Eternal* which was introduced about 1924. It was made in hard rubber and several years later, plastic models were available in a wide array of colors. A slightly less expensive pen in the Mabie Todd line was the Swallow Pen and one step down from that was the Blackbird Pen.

MONTBLANC 1908-Present

Montblanc's predecessor, the Simplo-Filler Pen Company was founded in 1908, in Germany. Simplo-Filler began by making eyedropper filled models. About 1920-1911, the company name was changed to the "Simplo Fountain Pen Company." Also at the time, Simplo began making a safety pen similar to the Waterman safety pen. The top of the line of the Simplo pens was called the **MONTBLANC**. It was made with a red or black barrel and cap and a crown of white. In 1913, the white cap evolved into the white star design which was registered as a trademark and looked more like the snow-topped summit of the Mont Blanc mountain which it was supposed to symbolize. Mont Blanc's height was 4,810 meters. This height eventually was marked on the MontBlanc pen nibs and indicated craftsmanship at its highest. In 1914 the company changed its name to "Simplo Fullhalter Gmbh." and began to expand its line of pens. In the 1920s, Simplo began making pens with ink sacs inside and a simple filling system similar to the Parker *Duofold* button filler. The "Masterpiece" pen, which later became the "Diplomat" was introduced in 1924. The MontBlanc trademark was so successful that, in 1934, Simplo Fountain Pen Company changed its name to the MontBlanc Pen Company. Also about this time, they introduced the twisting piston filler which is still made today. The MontBlanc pen evolved into the status symbol that it is today, but most collectors seek the earlier models. Many Mont Blanc pens were covered with precious metals whose design work is both beautiful and of the highest quality.

CROCKER PEN 1902-1931

CHILTON PEN 1923-1941

The Crocker Pen Company was started about 1902 in Massachusetts by Seth S. Crocker. Crocker Pen made very high quality eyedropper filled and blow filler pens. Crocker's first blow filler pen was patented July 16, 1901. It had a simple hole in the end of the barrel. One simply blew into it. The air collapsed the sac and the ink was drawn into the pen. Another blow filler was patented on August 9, 1904. This had the hole on the side of the end of the pen.

Seth S. Crocker's son, Seth Chilton Crocker, took over the company about 1907, renaming it the "S.C. Crocker Pen Company". He moved the company to 79 Nassau Street, New York City. Crocker made an improved blow filler pen and a Hatchet filler. His blow filler pen was patented on May 2, 1916. The name and patents of the Crocker Pen Company were sold to Nicholas Zaino in 1930-31.

In 1923, Seth C. Crocker started another pen company, the Chilton Pen Company, of Boston, Massachusetts and began using a new patented pump filler pen. These early Boston Chiltons are very nicely made pens. To fill the pen, the barrel would be pulled up from the section exposing a metal tube which contained an ink sac. Pushing down the barrel compressed the ink sac with air pressure. This design enabled the Chilton to hold a larger ink sac and more ink.

The company moved to New York in 1926, and apparently combined production of the Crocker and the Chilton lines. They began to make Chilton pens out of plastic in early 1926, which gave them access to new colors. In early 1927, the filler was slightly changed so that instead of the entire barrel moving up to pump, the end of the barrel was attached to a thick metal tube, with an air hold in the end, that slid out of the barrel which contained the ink sac.

Avery attractive line of Chilton pens, the *Wingflow* (1935-1938), was made in colored plastic with inset metal lines and initials. The Wingflow used a nib that was wrapped around and locked into the feed. The quality of Chilton pens declined over the next few years. Chilton Pen moved to Summit, New Jersey in 1939 and folded about a year-and-a-half later.

EAGLE PENCIL (and Pen) COMPANY
1860-Present

Eagle was founded in 1856 in Furth, Bavaria. When the founder brought his company and family to New York in 1860, they wanted an American symbol for the company so they chose the Eagle. The Eagle Pencil Company was located in New York City.

Their first successful, innovative pen was the glass cartridge pen of 1890. The pen came with a small glass vial filled with ink. The vial was inserted in the barrel and the section (nib, feed & collar) was connected to the ink vial with little mess. The body of the pen was either aluminum or painted brass. Another popular pen was their pen/pencil combination that most other pen makers produced in the 1930s. Eagle produced their first pen/pencil combination in 1890. They had been making dip pen/pencil combinations since 1870. This pen/pencil combination was simple black hard rubber.

By 1913, Eagle had a full line of fountain pens in different styles, sizes and shapes. By the 1920s, their pens reflected the styles of the pens of the day. They produced copies of most popular pens such as the Parker Big Red, Sheaffer Green Jade and Eversharp style pencils. About 1930, Eagle cut back its line of Eagle pens and began to produce the "EPENCO" line. The bright colors and clever designs of these plastic Epenco pens makes them a desirable addition to any collection. Epenco pens only cost 25 cents to a dollar when they were new.

CONKLIN PEN COMPANY 1898-1947

The Conklin Pen Company was founded by Roy Conklin in 1898 in Toledo, Ohio. Conklin is best known for their "Crescent Filler" pens (1901-1928). The Crescent Filler was patented in 1901 and 1903 and consisted of a bar inside the pen attached to a crescent that protruded outside the barrel which contained an ink sac. Ink was drawn into the pen by pushing down on the crescent. The crescent was locked in place by a knurled hard rubber band that turned freely around a groove in the barrel. The design was simple, no eyedropper was necessary and the pen enjoyed great success in sales. colors were usually black or mottle red & black with some models having silver or gold filled overlays. Conklin was among the four largest pen companies until Wahl took over that position in sales volume about 1924.

They made their popular *Endura* model from 1924 to 1932. They continued to make the crescent filler pens unchanged except for a squared off cap (about 1924). In 1930, they introduced a streamlined pen called the "Symetric" and in 1932 introduced their *Nozac* pen (1932-1938) designed by Louis Vavrik. It was available in a round or 10-sided version. The *Nozac* pen was different in that it had a twist filler similar to the MontBlanc filling system. Soje models had a word counter on the side of the transparent barrel that would tell the writer how many words worth of ink were left.

Conklin ran into financial difficulties during the Depression and sales lagged. In 1938, the company was sold to a syndicate in Chicago. This syndicate also made other pens: the Park-O-Type, the Waltham, the Winchester and the Starr Pen.

THE MOORE PEN COMPANY 1896-1956

THE AMERICAN FOUNTAIN PEN CO.
1899-1917

The Moore Pen Company was produced from a design by Mr. Moore, but Mr. Moore was never a part of the company. Moore patented a pen with a retractable nib on September 8, 1896. It was similar in design to the Waterman safety pen as the nib retracted into the barrel and was filled with an eyedropper. The pen was different in that a separate lower barrel was slid upward to push the nib straight up to writing position. To close the pen, the nib was withdrawn by sliding the lower barrel downward. The cap sealed the barrel and the pen did not leak. A partnership was soon entered into with Francis C. Brown to make the pen in New York City. The partnership ended in 1899 with Francis Brown getting the patents on the pen.

The first Moore pens were produced by the American Fountain Pen Company of 711 Boyleston Street, Boston. American Pen had acquired the patents to Moore's "Non-Leakable" pen and began producing the Non-Leakable line of Moore's Pens in late, 1899. The pen was a good seller and the company made a reasonable profit selling the same pen basically without modification until 1917. They changed their name and acquired some new pen patents and were reorganized as the Moore Pen Company.

Moore began making lever filler pens in 1918. They began to make plastic pens about 1927. In 1946, Moore attempted to produce a pen that would compete with the Parker *51* and the Sheaffer Triumpth. They created the "Fingertip Pen" (1946-1950) - a Buck Rogers-like, spaceship-shaped pen. It was an appealing pen that, when opened, looked as if it were from another world. Unfortunately for Moore, it was not a best seller. It makes a good addition to any collection of pens from the 1940s. Moore's last pen was the "Specialist" (1950-1956). It was a cheap, brass capped pen with an aerometric filler.

A.A. WATERMAN PEN CO. 1897-1920

The A. A. Waterman Pen Company was founded about 1897 in New York City by Arthur A. Waterman (no relation to L. E. Waterman). A. A. Waterman was known for their New Lincoln pen (1904-1907) which was a regular eyedropper filler pen, their middle joint filler pen (1902-1907) and their "Modern Automatic Self Filler", a twist filler (1902-1920). There are some exceptionally beautiful models made in gold filled metal and silver. About 1912, L. E. Waterman sued the A. A. Waterman Company and forced them to put a disclaimer stating, "Not connected with the L. E. Waterman Co." on all of their pens and advertising. The A. A. Waterman Pen Company was gone by the 1920s.

JOHN HOLLAND PEN CO. 1841-1950

The John Holland Pen Co. of Cincinnati, Ohio was one of American's oldest pen makers. The predecessor company began making gold nibs for dip pens in 1841. The company was incorporated in 1885.

Holland pens are good quality and rather rare. By 1920, all of their pens were lever fillers. Holland made the "MacKinnon Pen" which was a very popular 1878-1880s stylographic pen. Pen names used by Holland were "Clymax", "Columbia", "Dexter", "Duplex", "Durand & Co.", "Fountgraph", "Geo. W. Sheppard", "W. B. Snow", "MacKinnon", "Victor" and "Imperator".

DUNN PEN COMPANY 1921-1924

One of the shortest lived companies, Dunn Pen, was founded in 1921 in New York City. The Dunn was a pump filler and was identifiable by its red section at the back end of the barrel. The pen held a great deal of ink due to its filling system. Quality was high but few of these pens work today due to the hardening of the rubber pump parts. Dunn went out of business in 1924 after a prolonged strike by its employees.

HOUSTON PEN COMPANY 1911-1924

The Houston Pen Company of Sioux City, Iowa, was founded by William A. Houston. The style of the pen set it apart from other pens. It looked like a desk pen with a lady's style cap (no clip) with a safety chain attached. Filling systems were an eyedropper filler and a pump filler which was replaced by an unusual, modified matchstick filler system about 1915-1916. Houston also made the "Jiffy Pen Co." pump filler pen of the same style. Houston Pen became the General Manufacturing Co. in 1920 which manufactured the "Snapfil Fountain Pen".

THE LINCOLN PEN CO. 1895-1906

The Lincoln pen was first introduced in 1895. It was a simple unassuming pen that was very well made, cost one dollar, and had a gold nib. Lincoln Pen was located in New York. In 1899 it was sold to the Frazier & Geyer Company. F & G expanded the Lincoln line to include different priced pens, including a solid gold overlay pen. In 1904, they entered into a partnership with the A. A. Waterman Pen Company and began to sell the Lincoln and later, the "New Lincoln" pen, under the A. A. Waterman banner. The New Lincoln was A. A. Waterman's standard eyedropper pen line. Apparently after several uneventful selling years, the New Lincoln was lost to obscurity. A Lincoln pen resurfaced in the 1920s made by the National Pen Company of Chicago, Illinois. The 1920s pen was comparable to an Eclipse or Gold Medal pen.

CARTER PEN CO. 1926-1931

John W. Carter founded a company in Boston, Massachusetts in 1857 that would later become the Carter Ink Company. Carter was best known for its ink products, but made very nice fountain pens during the 1920s and 1930s. Fountain pen manufacturing began about 1926. Carter pens appeared in some very unusual colors. These included shades of green, blue and orange and an white pearl. Carter abandoned the pen business about 1931 and returned to jest selling ink products.

AIKIN LAMBERT PEN CO. 1864-1932

The Aikin Lambert Company was begun in 1864 in New York City. It initially made gold pen points for dip pens and began to make fountain pens in the 1890s. The majority of Aikin Lambert's pens were high quality, with many of them having beautiful silver and gold filled overlays similar to pens made by L. E. Waterman. Also, many styles of pens were made including Waterman safety style, eyedroppers and lever fillers. Few know that L. E. Waterman's pencils were made by Aikin Lambert until about 1920. They merged with Waterman Pen Company about 1932.

PAUL E. WIRT FOUNTAIN PEN CO. 1878-1930

The Paul E. Wirt Fountain Pen Company was started by Paul E. Wirt in Bloomsburg, Pennsylvania. Wirt patented his first pen design in 1878 and made a few pens by hand. He went on to open his first factory in 1885. Wirt made a popular, simple and reliable pen that was picked up by several major retailers, including Sears & Roebuck. Wirts were handsome pens and examples have been found in heavy silver and gold filled overlays. Sales began to slip by the early 1920s. The company finally closed about 1930.

OTHER PEN COMPANIES

There were hundreds of pen companies. Some lasted only a few years while others made millions of pens that were sold through mail order catalogs or sold in only the regional area. Quality varies greatly. They are collectible but the number of collectors seeking these lesser known brands is substantially less that those seeking pens made by the larger, high quality pen companies.

Valuing Fountain Pens

Valuing older fountain pens can be approached several ways. The easiest method for a new collector is to compare the price of an older pen with a modern pen. The older pen should be valued at least equally with the new pen. Pen collectors, however, use a "Supply and Demand" valuation or "what would a willing buyer and a willing seller agree to as a value?" An examination of the components of value are helpful. A valuable pen is usually rare, but a rare pen is not always valuable. A rare pen, of which only one or two are known may not have the broad appeal of say the 1906 Parker "Snake" pen. There are more than 40 Parker Snakes in collections and yet the knowledgeable collector would jump at the opportunity to pick one up at a bargain price of $5,000. A 1940s Superman pen, of which there are only 2 known in collections, may sell for less than $250. Obviously, the rarity factor is not the main determinant of value. It depends upon who is buying, the availability and how badly he wants that pen.

Things to consider when valuing Fountain pens:

Condition is the most important criteria. A Parker *Duofold* Deluxe, in mint condition, may be worth 5 times one in extra fine condition. Damage to a cap or barrel is a major problem. A pen with a small crack in the cap or barrel may be worth one quarter of one without a crack. Repair is often possible, but to repair the pen, one needs parts or another pen with that part. Ask yourself if the price is still a bargain when you have to cannibalize another pen for parts.

Color: Tens of millions of pens were made in black, less were made in color. The quality of the color is also important. A perfect pen in a crisp, rich color may demand a premium over a standard, well-colored pen. The pens illustrated show good examples of color.

Working Condition: Pens should be in working condition if possible. While the collector who can put a pen in working order may not care if the pen needs cleaning and a new ink sac, most collectors would rather not do that job. During cleaning, the pen can be broken. If you are not sure what you are doing, have someone else put your pens in working order. Ideally you will find pens that have been repaired and are in working condition (new ink sac & cleaned). Figure $20 for cleaning and new ink sac.

Size: Collectors usually look for the largest versions of a pen. Lady's pens, the smaller pens with the ring on top, have remained undervalued. May of these pens rival the men's pens for beauty and workmanship and certain colors are available only in lady's pens. Presently, more men than women collect fountain pens. Men want a pen that can be kept in a jacket or shirt pocket. Lady's pens were designed to be worn on a ribbon or kept in a pocket book. Good values abound in quality Lady's pens.

The standard size pen in 5 1/4" (13.4cm) to 5 1/2" (14 cm) long when closed. Often the highest priced pens are the largest men's pens, all over 5 1/2" (14 cm) long when closed. Examples are the Waterman #20 and #58, the Parker Black or Red Giant, large *Duofolds*, Oversize *Vacumatics*, full sized Sheaffers and Wahl-Eversharps. They can be compared with the most popular modern selling pen - the large Mont Blanc Diplomat. Many collectors simply want the biggest, high quality pen available.

Original Parts: Not all pens need to have original parts but they are very important to some people, especially Parker *Duofold* collectors. Parker *Duofolds* were originally made with the stiff *Doufold* nib and the "Christmas tree" feed. The later *Vacumatic* nib and feed were more flexible and wrote differently from the *Duofold* nib and feed. During the 1930s, pen stores often replaced the *Duofold* nib and feed with the *Vacumatic* nib and feed when making repairs. Also in the 1930s, Parker sold its remaining *Duofold* stock with the *Vacumatic* nib & feed. Although technically correct on the 1930s *Duofold* pens, a *Duofold* pen with *Duofold* nib and feed may sell for 25% more than the same pen with a *Vacumatic* nib & feed.

Early Pens - 1910 and before: Collectors are looking for the earliest pens made by the major pen manufacturers. Rarely do they expect to use them. Nevertheless, they should be in working condition to maintain value. When made by Waterman or Parker, they early pens, expecially silver or gold covered, are more valuable than many of the later pens. As pen collecting has matured, collectors have rediscovered the early pen makers.

Novel Filling Systems: These normal-sized, black pens often contain ingenious methods for making the pen more leakproof, easier to fill, or able to hold more ink. Some manufacturers were just trying to avoid patent infringment lawsuits while others thought they had a better idea. For example, Crocker, Sanford and a few other makers designed "Blow filler" pens. You put the cap on the back end of the pen and blew into it. The air pressure compressed the sac and when you stopped blowing, the sac expanded drawing ink into the pen. Many innovative makers' pens are very collectible and can often be obtained at bargain prices.

Price: When deciding whether to buy a pen, remember that in a few years you will rarely regret paying too much for a pen that you own, but you will always regret having not bought a good pen when it was available.

IMPORTANT NOTICE:

1) The pens illustrated vary in condition. Most are excellent to near mint. The prices are based upon a pen in excellent condition that has been cleaned and has had an ink sac installed where applicable. Add 20-25% if in mint condition (with Parker *Duofolds*, quality of the color may vary the price even more so). Subtract about $20.00 if the pen has not been cleaned or repaired. Subtract 50-75% if the pen is missing any parts. Subtract 20% if the condition is extra fine, 30% if in fine condition and 50% if in only good condition. These are retail prices. If you are trying to sell your pen to a pen dealer, expect 50% or so of the above figured value to be a reasonable offer. Do not fool yourself into believing that a defect or flaw will not affect value. Look over the pen with a magnifying glass, most dealers will do the same.

2) The fact that a pen is not illustrated here does not mean it is rare or valuable. It may be extremely common. This book is a guide and at the time of preparation, these were the pens that were available to illustrate it. For a more comprehensive guide to pens, read *Fountain Pens and Pencils - The Golden Age Of Writing Instruments and The Book Of Fountain Pens*. Both are large format, hard cover books with over 1800 color photographs of fountain pens, pencils and pen company advertising produced by the same authors as this book.

Pens

1905 A.A. WATERMAN red & black woodgrain hard rubber, eyedropper filler, taper cap pen. $195-235

1912 AIKIN LAMBERT pen, gold filled hand-engraved design overlay, eyedropper filler. $395-450

1905 AIKIN LAMBERT gold filled chased design half overlay, taper cap, eyedropper filler pen. A very nice, rare pen possibly made by Waterman for Aikin Lambert. $385-445

1927 AIKIN LAMBERT full size "Mercantile" pen, perfect bright jade green, gold filled trim, lever filler. $125-150

1932 BELMONT pen/pencil combo, red veined grey marble. $30-40

1950 BAYARD 2000, maroon, gold-filled trim, lever filler. $65-80

1950 BERMOND, black, chrome cap, twist filler. $20-30

1929 BLUE JAY pen made by Blackwell-Weilandy B & S Co., St. Louis, blue, lever filler. $85-120

1937 CARAN D'ACHE pencil. $35-45

1914 BOSTON pen, black, lever filler, gold-filled trim. $85-100

1910 CAREY pen, black, eyedropper filler, no markings on barrel. $50-60

1931 CARTER Inx pen, black, gold-filled trim, lever filler. $200-250

1895 CAW'S "stylographic" pen, red & black hard rubber, eyedropper filler. Practically the same pen as the early A. T. Cross but much rarer. Caws was a short lived, high quality pen producer. $145-160

1982 CARTIER pencil, sterling silver. $30-40

1927 CHATSWORTH pen,
English made, lever filler, nickel
plated trim, black chased hard
rubber. $45-55

1918 CHICKERING pen, black,
eyedropper filler, from the Boston
Piano manufacturer. $75-100

1935 COLUMBUS EXTRA
pen, blue pearl, gold-filled
trim, bar filler under blind cap.
This Italian pen has a Parker
style clip, an Eversharp style
body & hand, and is very well
made. $300-350

1936 CHILTON holy water
sprinkler, black with silver
cross, pump filler. $300-350

888 COMFORT pen, black chased hard rubber, eyedropper filler, overfeed. $45-60

1919 CONKLIN crescent filling #25P pen, black chased hard rubber, gold filled trim. $45-60

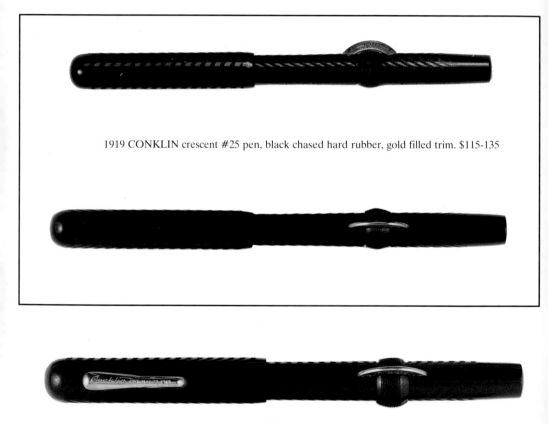

1919 CONKLIN crescent #25 pen, black chased hard rubber, gold filled trim. $115-135

1919 CONKLIN Large crescent filling #40, black chased hard rubber, gold filled trim. $200-250

1919 CONKLIN crescent filling pen in gold filled overlay with stars. $1500-1750

1920 CONKLIN gold-filled hand-engraved overlay, #3 nib, one of the nicest large Conklin "Crescent" filler pens. $1000-1250

1925 CONKLIN pen, orange hard rubber, lever filler, gold filled trim. $90-125

1925 CONKLIN pen, gold filled, lever filler. $75-110

1925 CONKLIN pen, white rolled gold (gold filled). $50-75

1925 CONKLIN *Endura* pen, black, gold filled trim, lever filler. $45-60

1927 CONKLIN pen, black chased hard rubber, lever filler. $45-60

1929 CONKLIN pen, black chased hard rubber, lever filler. $85-120

1929 CONKLIN *Endura* pen, lapis blue, lever filler. $175-225

1930 CONKLIN *Endura* lapis blue with gold filled trim pencil. $35-45

1930 CONKLIN *Endura* pen, gold & black marble, gold filled trim, lever filler. $155-180

1930 CONKLIN *Endura* pen, black, gold filled trim, lever filler. $145-160

1932 CONKLIN "*Endura*" pen, black veined burgundy & grey pearl, lever filler, gold filled trim. $75-115

1932 CONKLIN "*Endura*" pen, black veined green pearl, lever filler, gold filled trim. $185-225

1932 CONKLIN Stylographic pen, red specked grey pearl, lever filler, gold filled trim. $150-175

1933 CONKLIN desk pen, black with orange taper, gold filled trim, lever filler. $65-85

1940 CONWAY STEWART #58, grey pearl stripe, gold filled trim, lever filler. $105-135

1916 CROCKER hatchet filler pen, black chased hard rubber, gold filled trim. $75-90

1950 CONWAY STEWART #60 Duro, black, gold filled trim, lever filler. $150-175

1890 CROSS or CAW (no markings) stylographic pen, red & black hard rubber. $125-150

1927 DIAMOND POINT Emblem pen, black, gold filled trim, lever filler, large #8 nib. $175-245

1922 DUNN baby pen, black chased hard rubber, gold filled trim. $75-100

1927 DIAMOND POINT pen, red & black Rosewood with black ends, lever filler, gold filled trim. $155-200

DIAMOND POINT, ca. 1932-1935. The Babe Ruth pen, available only in ballparks and actively sought by baseball item collectors Two styles, one with Babe's name on the clip and the other with his name on the barrel. $300-450

1926 DUNN pen, full-size black pump filler, large #6 nib. $175-225

1919 EAGLE pen in gold plated metal, lever filler. $50-65

1921 EAGLE "Russet" pen in red hard rubber, stylographic. $35-40

1937 EAGLE "Epenco Merlin" pen, red with gold stars & moons, lever filler, gold plated trim. $35-50

1937 EAGLE "Epenco Merlin" pen, brown with gold stars & moons, lever filler, gold plated trim. $35-50

1937 EAGLE "Epenco Merlin" pen, black with gold stars & moons, lever filler, gold plated trim. $35-50

1927 ECLIPSE pen & pencil set in gold filled metal, lever filler. $165-200

1900 ECKFELDT & ACKLEY sterling silver long mechanical pencil Eckfeldt & Ackley were silversmiths in Newark, New Jersey from 1896-1934. $45-55

1937 EAGLE "Epenco Merlin" pen, green with gold stars & moons, lever filler, gold plated trim. $35-50

45

1927 ECLIPSE pen & pencil
set in grey plastic, lever filler.
$40-50

1926 EDWARD TODD,
14kt heavy gold, lined
pattern, lever filler.
$450-525

1932 ECLIPSE
"Gothic" pattern pen,
gold filled, lever filler.
$85-110

1926 EDWARD TODD pen, 14kt
heavy gold with line design, lever
filler. $300-350

1928 EDWARD TODD, sterling silver herring bone design, lever filler. $295-345

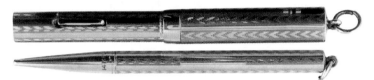

1929 EDWARD TODD pen/pencil combination, lever filler, gold filled. $200-250

1932 EDWARD TODD pen & pencil set in gold filled metal. $150-175

1932 EDWARD TODD pen/pencil combo in sterling silver ring pattern, lever filler. $250-280

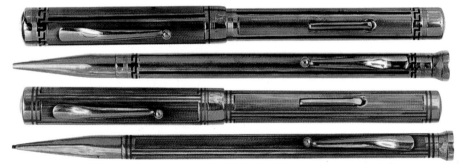

EDWARD TODD, ca. 1928. Two pen and pencil sets by this highest quality pen maker. Both sets are 14kt gold with enamel work. These are truely elegant pens and pencils. $1500-2100 each set.

EDWARD TODD, ca. 1928. Six sided 14kt gold mechanical pencil with a machine engraved design. Edward Todd's hallmark is a sideways rounded "E" with a "T" coming out of the center line of the "E". $125-150

EDWARD TODD DUPLEX, ca. 1930. Black with gold veining pen/pencil combination by this top quality maker. $300-400

47

1929 EMPIRE "Special Complete" pen, black chased hard rubber, nickel plated trim, lever filler. $35-55

1930 EVEREADY PEN, yellow, lever filler, 14kt point. $45-55

1932 FRANKLIN pen/ pencil combination, jade green. Franklin was a high quality pen company in Philadelphia, rare, gold filled trim, lever filler. $250-300

1952 ESTERBROOK nurse's set, white with nickel plated trim, lever filler. $100-125

1960 GEHA, black, gold filled trim, made in Germany, twist filler. $45-60

1921 GENERAL MFG CO. "Snapfil" pen, black, lever filler. $85-110

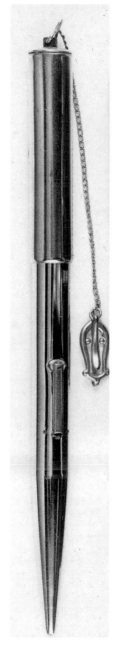

1932 GEORG JENSEN pen/pencil combo by Hutcheon Bros., sterling silver, lever filler. $300-350

1919 GENERAL MFG CO. "Snapfil" pen in all gold filled metal, Snapfil lever filler, with chain & pin, RARE in metal versions. See page 51 of *The Book of Fountain Pens and Pencils*. $350-400

1910 GESSNER pen, gold filled, made in New Orleans, Louisiana, eyedropper filler. $95-120

1927 GOLD METAL pen, red streaked grey pearl, lever filler. $50-60

1928 GOLD BOND, large size, lapis blue with yellow ends, gold filled trim, lever filler. $175-210

1927 GOLD BOND pen, green pearl, lever filler. $55-65

1952 GOLDEN KING pen, gold filled, unusual pump filler, 14kt gold nib. $35-45

1926 GREISHABER pen, green, pink & gold handpainted on black, lever filler. $250-300

HAMILTON ROSS, ca. 1947. This handsome "Jet Flow" ball point pen, made in Chicago, was one of the earliest ball points. The pen was made of anodized aluminum and originally sold for $10. It used a twist mechanism to retract and extend the point. $70-90

1925 JOHN HANCOCK pen with its original box with a triple cartridge pack, black, early cartridge filler, made in Boston, gold filled trim. $125-155

1927 HARTLINE pen, red, wire twist filling, with rolling ceramic blotter under the blind cap in the crown. Made in Tampa, Florida. See page 55 of *The Book of Fountain Pens and Pencils* for additional information. $265-300

1910 HEATH sterling silver overlay pen, an exact duplicate of a Parker #14. Heath made the overlays for Parker pen, has a Parker nib and Lucky Curve feed, eyedropper filler. $800-950

1910 HEATH sterling silver overlay pencil holder, unusual piece that went with their #14 pen, rare. $175-200

1935 HICKS pen and pencil, sterling silver, lever filler. $245-300

1932 HICKS pen/pencil combination in 14kt gold, barleycorn pattern, lever filler. $375-450

1918 HICKS bullet pen in 14kt smooth gold, eyedropper filler, super pen only 3.5" long. $300-325

1885 JOHN HOLLAND dip pen/pencil, maker only marked on nib, gold filled trim. $35-45

1925 JOHN HOLLAND
pen, orange hard rubber,
gold filled trim, lever filler.
$195-225

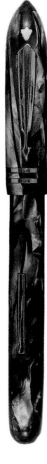

1940 JOHN HOLLAND
pen, black-veined grey
pearl, gold filled trim, lever
filler. $100-130

1928 JOHN HOL-
LAND "Jewel" pen,
pearl & black, gold
filled trim, lever filler.
$175-220

1939 KINGSWOOD pen,
burgundy pearl marble,
made in England, gold
filled trim. $40-50

1933 INDIAN pen/ pencil combo, red, black, white & green, nickel plated trim, lever filler. $60-90

1908 HUTCHEON "Clutch" pencil in hand-engraved sterling silver. $120-160

1933 INDIAN pen/ pencil combination, green with red, white & black design, lever filler, nickel plated trim. $60-90

1933 INDIAN pen/ pencil combination, blue with red, white & green design, lever filler, nickel plated trim. $60-90

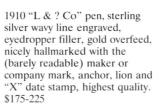

1910 "L & ? Co" pen, sterling silver wavy line engraved, eyedropper filler, gold overfeed, nicely hallmarked with the (barely readable) maker or company mark, anchor, lion and "X" date stamp, highest quality. $175-225

1925 INKOGRAPH full size stylographic pen, black, nickel plated trim. $65-95

1937 INKOGRAPH pen, blue pearl marble, lever filler, gold plated trim. $20-30

1926 LAKESIDE pen, full size in red hard rubber, gold filled trim, see page 64 of *The Book of Fountain Pens and Pencils*. $125-150

1962 LAMY 2000 pen &
ballpoint, black & chrome,
cartridge filler. $75-100

1964 LAMY 25P pen, black & chrome,
cartridge filler. $40-60

1915 LAUGHLIN stylographic pen, black,
eyedropper filler, interesting air admitting
feature. $65-90

1932 LEBOEUF Desk pen and base in two tone silver pearl, lever filler, gold filled trim. $375-450

1932 LEBOUEUF pen,
green grey cat's eye
pearl, lever filler. $195-
225

1925 LEONARDO pen, gold
filled with wave design, lever
filler, 14kt nib. $20-30

1946 LIBERATOR, black,
made in France, gold filled
trim, button filler. $30-40

1928 LINCOLN by National Pen
of Chicago, full size, red & black
hard rubber, gold filled trim, lever
filler. $175-200

1917 MASTERPEN,
black chased hard
rubber. Masterpen
was made by Julius
Schnell, nickel plated
trim. $70-90

1950 LUXOR, black,
gold filled trim, made in
Germany, twist filler.
$50-80

1889 McKINNON
pen, eyedropper
filler, black chased
hard rubber, hand-
engraved gold filled
trim. $200-225

1932 MONROE pen,
beautiful black veined
pearl, lever filler, gold
filled trim. $95-145

1950 MERCURY student,
grey pearl, gold filled
trim, made in Belgium,
twist filler. $35-45

1930s MONTBLANC pocket pen holder for two pens, green leather. $50-70

1930-40 MONTBLANC pocket pen holder, red leather. $30-35

1950 MONTBLANC #212 *"Duofold"* red with gold filled trim. MontBlancs in red are among the most difficult to find and this is the rare button filler model. $225-300

61

1950 MONTBLANC Monte
Rosa 042G pen, black, gold
filled trim, twist fill. $100-130

1970 MONTBLANC 224 pen,
black with brushed gold filled cap,
twist fill. $125-160

1970 MONTBLANC Carrera, black,
brushed chrome cap, cartridge. $40-50

1970 MONTBLANC Carrera fountain pen & ballpoint set, yellow, black & chrome cap,
cartridge. $95-135

1975 MONTBLANC 149 "Masterpiece", full size, black, gold filled trim, pump filler, large 18kt point, two tone nib. $195-225

1908 MOONEY pen, black hard rubber blowfiller, marked "The Mooney Model Fountain Pen, Hauk & Gross Mfrs, Chicago". $150-180

MOONEY, ca. 1900-1910. A sterling silver "Filigree" design overlay blow filler pen. The cap was fitted on the rear end of the pen and the user blew into the end. This compressed the ink sac and allowed it to draw in ink as it decompressed. $750-1000

1940 MOORE 94-A "Lifetime" pen, black, gold filled trim, lever filler. $55-75

1944 MOORE pen, black, lever filler, gold filled trim. $65-80

1925 MORRISON pen, gold filled with wave design, lever filler, 14kt gold nib. $60-85

1946 MOORE "Fingertip" pen, cream color, gold filled trim, lever filler. $175-235

1946 MOORE "Finger-tip" pen, black veined silver pearl, rare & beautiful, gold filled trim, lever filler. $400-500

1946 MOORE "Fingertip" pen, black veined gold pearl, rare & beautiful, gold filled trim, lever filler. $400-500

1942 MORRISON "The Patriot" Army
pen, khaki green, pump filler, 14kt gold
nib. $70-85

1946 NORMA four color pencil,
gold filled. $30-50

1919 ONOTO "Ink pencil", black
with 18kt gold bands. $100-135

1920 ONOTO stylographic "Ink
pencil" pen, black, gold filled trim,
eyedropper filler. $60-80

1914 PARKER #28 1/2 jack knife safety pen, black hard rubber, nice large #8 nib. The predecessor to the Senior black *Duofold* without band, nickel plated trim. $300-400

1909 PARKER #42 1/2 pen, black hard rubber, eyedropper filler, gold filled trim. $650-800

1910 PARKER #16, gold filled overlay pen. Heath made the overlays for Parker pen, very rare, eyedropper filler. $950-1100

1918 PARKER #18 Lucky Curve pen, black hard rubber, eyedropper filler. $125-145

66

1922 PARKER *Duofold* red pencil, gold filled trim. $40-50

1919 PARKER Jack Knife Safety pen, black chased hard rubber. $55-75

1920 PARKER #20 Lucky Curve pen, black hard rubber, nickel plated trim. $85-110

1920 PARKER #22 1/2 Lucky Curve pen, black chased hard rubber, blowfiller. $85-110

1924 PARKER *Lady Duofold* red pen, gold filled trim, original feed & nib. $65-90

1924 PARKER *Duofold* Jr., red, original nib & feed, gold filled trim. $95-135

1926 PARKER *Duofold* Sr., red, original nib & feed, gold filled trim. $250-325

1927 PARKER *Lady Duofold*, black, gold filled trim. $40-50

1928 PARKER "Pastel", coral (salmon colored), gold filled trim. $80-100

1928 PARKER *Duofold* Sr., red, later feed & nib, gold filled trim. $175-220

1928 PARKER *Duofold* Jr., black veined pearl, original nib & feed. $100-125

1929 PARKER *Lady Duofold*, black veined pearl, original nib & feed. The color is near perfect. $80-100

1929 PARKER *Lady Duofold*, red, original nib & feed, gold filled trim. $65-75

1929 PARKER *Duofold* Sr. Deluxe in an incredible brown hued sea green pearl, gold filled trim, original *Duofold* nib & feed, rare color large *Duofold*. $400-500

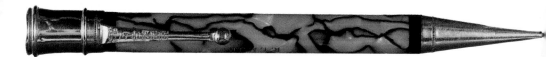

1929 PARKER *Duofold* Sr. pencil, color is a beautiful even yellow pearl with black veining, gold filled trim. $90-110

1929 PARKER pastel pen Naples blue, gold filled trim. $100-135

1930 PARKER True Blue. $125-175

1930 PARKER *Duofold* Sr., black, original feed & nib, gold filled trim. $175-250

1930 PARKER *Duofold* Jr., black, original nib, later feed, gold filled trim. $75-125

1930 PARKER pencil in rare black & gold, gold filled trim. $65-110

1930 PARKER *Lady Duofold*, green veined pearl, original nib & feed. $65-85

1930 PARKER *Lady Duofold*, black veined pearl, original nib & feed. Color is perfect. $125-145

1931 PARKER *Duofold* Sr. Streamlined Deluxe, black veined pearl, gold filled trim. $275-400

1932 PARKER "Vacuum-Filler", black and amber see-thru barrel (demonstrator), the first *Vacumatics* were known as "Vacuum-Fillers". The barrel and nib are marked, "Vacuum Filler", gold filled trim. $375-500

1932 PARKER moire coral pencil, gold filled trim. $45-55

1933 PARKER *Duofold* Sr. in purple-hued grey pearl, gold filled trim, original *Duofold* nib & comb fee. $500-750

1933 PARKER *Duofold* "Special" in a wonderful blue pearl marble, gold filled trim. $300-400

1933 PARKER *Duofold* "Special" in a wonderful red pearl marble, gold filled trim. $300-400

1935 PARKER Full size *Duofold*, green pearl toothbrush design, gold filled trim. $250-275

1935 PARKER Depression era pen, black, gold filled trim. $75-100

1935 PARKER standard
size *Duofold*, pen & pencil,
grey pearl, toothbrush
design, nickel plated trim.
$250-275

1935 PARKER *Deluxe
Challenger*, grey pearl, nickel
plated trim, button filler.
$45-65

1935 PARKER Chal-
lenger, grey pearl, nickel
plated trim, button filler.
$40-60

1935 PARKER three band
Standard *Vacumatic*, gold
pearl, gold filled trim, two tone
point. $75-100

1936 PARKER *Royal Chal-
lenger*, red pearl with sword clip,
gold filled trim. $250-300

1936 PARKER oversize
Vacumatic, black, Canadian
made, gold filled trim. $300-350

1936 PARKER *Deluxe Challenger*, green pearl, blowfiller, gold filled trim. $55-85

1936 PARKER Standard *Vacumatic*, golden pearl, gold filled trim. $75-100

1936 PARKER three band Standard *Vacumatic*, grey pearl, nickel plated trim. $75-100

1937 PARKER two band Slender Jr. *Vacumatic*, burgundy pearl, gold filled trim. $100-110

1938 PARKER "*Duofold*" English-made Vacumatic styling with *Duofold* nib & filler, black, gold filled trim. $100-125

1938 PARKER (larger size) *Royal Challenger*, silver pearl, nickel plated trim. $225-275

1938 PARKER "*Duofold*", English-made Vacumatic styling with *Duofold* nib & filler, brown, gold filled trim. $100-125

1938 PARKER *Royal Challenger*, red pearl with step clip, gold filled trim. $225-275

1939 PARKER *Challenger*, blue pearl, gold filled trim. $65-80

1939 PARKER standard size Blue Diamond *Vacumatic*, black, gold filled trim. $70-90

1940 PARKER two band Standard *Vacumatic*, grey pearl, nickel plated trim. $65-80

1939 PARKER full size "Maxima" *Vacumatic*, emerald green pearl, gold filled trim. $300-400

1940 PARKER unusual lined wide band
Standard *Vacumatic*, grey pearl, nickel
plated trim. $70-90

1941 PARKER *Duofold*,
burgundy & grey pearl, gold
filled trim. $65-80

1941 PARKER Blue
Diamond Debutante
Vacumatic, grey pearl,
nickel plated trim. $65-75

1941 PARKER *Duofold*,
blue & grey pearl. $75-90

1941 PARKER *Duofold*,
black striped green & gold
pearl. $65-90

1941 PARKER brown
stripe Shadow Wave
Vacumatic, gold filled
trim. $150-225

1941 PARKER Blue
Diamond *Duofold*, blue
pearl, gold filled trim.
$90-135

1941 PARKER *Duofold*, blue
pearl, gold filled trim. $50-60

1945 PARKER Blue
Diamond Standard
Vacumatic, gold pearl,
gold filled trim. $75-90

1941 PARKER Debutante,
green/brown pearl, gold
filled trim. $75-90

1944 PARKER *51* pen &
pencil, black with sterling
silver caps. $200-250

1945 PARKER Blue
Diamond "51", black with
gold filled cap, Beautiful
tapered engraved line
design on cap. $95-145

1945 PARKER wide band Standard *Vacumatic*, green pearl, gold filled trim. $75-90

1945 PARKER Blue
Diamond *Vacumatic*,
green pearl, gold filled
trim. 475-100

1945 PARKER "*51*"
pen, black with gold
filled cap & 14kt gold
clip. $150-175

1946 PARKER *51* pen,
burgundy with sterling
silver cap, gold filled
trim. $100-125

1946 PARKER *51* pen, black with 14kt gold cap, clip and lower ring. $500-650

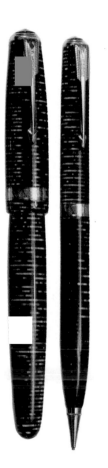

1946 PARKER Blue Diamond
Standard *Vacumatic* pen &
pencil, gold pearl, gold filled
trim. $125-145

1947 PARKER *51*, dark
blue with brushed
chrome cap. $45-60

1950 PARKER *51*,
medium blue with
brushed chrome cap,
Aerometric filler. $45-60

1950 PARKER *51* Demi, dark green with chrome cap, Aerometric filler. $50-75

1950 PARKER *51* Demi pen, black with gold filled caps, Aerometric filler. $75-90

1950 PARKER *41*, a light robin's egg blue/grey, brushed chrome cap, made for only one year. $75-100

1950 PARKER *41*, light pink, brushed chrome cap, made for one year. $75-100

1954 PARKER Jotter ballpoint pen, first model with grooved nylon barrel and no metal end piece, black. $45-65

1959 PARKER Jotter vase top ballpoint pen, light blue with chrome plated trim. This is a "Proto-type" and was not commercially marketed, but made up as a sample to test customer interest. Interest was mild. $45-65

1959 PARKER Jotter vase top ballpoint pen black with chrome plated trim, prototype. $45-65

1959 PARKER *61*, gold filled pen & pencil, lined design. $180-225

1959 PARKER *61*, turquoise blue with Lustraloy cap. $75-95

1960 PARKER *45* Demonstrator pen, black with chrome plated trim, marked "Demonstrator". $25-30

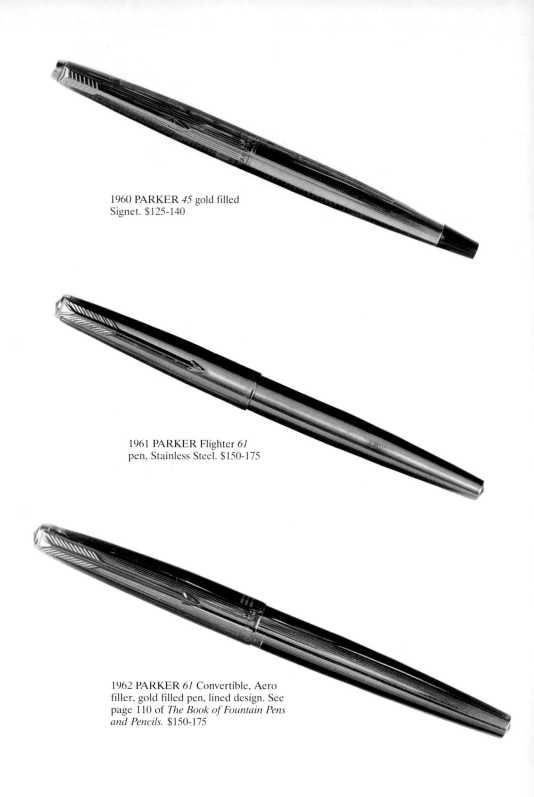

1960 PARKER *45* gold filled
Signet. $125-140

1961 PARKER Flighter *61*
pen, Stainless Steel. $150-175

1962 PARKER *61* Convertible, Aero
filler, gold filled pen, lined design. See
page 110 of *The Book of Fountain Pens
and Pencils.* $150-175

1970 PARKER *T-1* fountain pen, titanium with gold filled trim. The T-1s were made for only one year and are incredibly popular, adjustable point. $500-600

1970 PARKER *T-1* felt tip pen, titanium with gold filled trim, actually rarer than the fountain pen. $250-300

1970 PARKER *T-1* ballpoint pen, titanium with gold filled trim, also rarer than the fountain pen, takes a Parker Jotter refill. $250-350

1975 PARKER *75* Flighter pen, brushed steel. $155-225

1980 PARKER *180*, gold filled. $75-100

1952 PELIKAN *400* pen, green & black stripe, pump filler, gold filled trim. $85-100

1991 PARKER with sterling silver overlay, burgundy, a custom pen remade in a limited edition, the filigree overlay is heavy silver. $275-400

1952 PELIKAN pencil, green &
black stripe, gold filled trim,
mint in box with instructions.
$70-80

1954 PELIKAN 400 pen, brown &
brown stripe, gold filled trim,
pump filler. $125-150

1965 PELIKAN MK 30,
black, gold filled trim, twist
filler. $40-50

1965 PELIKAN MK 20 silver,
black, chrome cap, twist filler.
$40-50

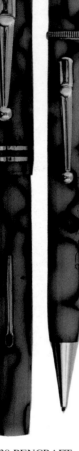

1930 PENCRAFT pen & pencil, black veined pearl, gold filled trim. $100-125

1933 PICK Tuckaway pen, cream & black, gold filled trim, lever filler. $75-90

1933 POSTAL full size pencil blue/green veined perfect pearl. The style and color of this pencil is most unusual, gold filled trim. $75-95

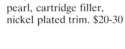

1980s RECIFE pen, full size, beautiful brown pearl, cartridge filler, nickel plated trim. $20-30

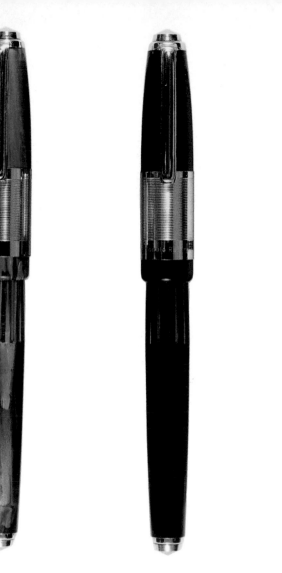

1980s RECIFE pen, full size, beautiful green pearl, cartridge filler, nickel plated trim. $20-30

1980s RECIFE pen, full size, black, cartridge filler, nickel plated trim. $15-20

1924 SALZ black & white, black with white ends, lever filler, nickel plated trim. $35-40

1926 SALZ BROS. pen, sterling silver with filigree design, lever filler. $165-200

1920 PETER PAN by SALZ pen, 2.25
inches long, hand-engraved gold filled,
eyedropper filler. $200-250

1925 PETER PAN by SALZ pen, gold filled, lever
filler. $75-90

1912 SANFORD &
BENNETT pen,
alternating pearl &
abalone slabs, gold
filled trim, eyedropper
filler. $275-350

1922 SECURITY pen,
black chased hard
rubber, gold filled
trim. $150-200

1913 SHEAFFER #34 early pen with 1912 patent date, lever filler, early feed and early #4 nib. $90-110

1918 SHEAFFER, #2A, black chased hard rubber, nickel plated trim, lever filler. $75-90

1920 SHEAFFER gold filled pen & pencil, lever filler, initials. $145-170

1923 SHEAFFER pen & pencil, tiny gold filled, lever filler. $145-175

1923 SHEAFFER, 46 Special, black chased hard rubber, gold filled trim, lever filler. $65-80

1923 SHEAFFER gold filled pen, wavy lined, lever filler. $100-125

1925 SHEAFFER, *Lifetime* cherry red full size pencil, gold filled trim. $135-165

1926 SHEAFFER full size light jade green *Lifetime* pen, gold filled trim, lever filler $150-250

1925 SHEAFFER, "Secretary" cherry red, full size pen, gold filled trim, lever filler. $450-550

1928 SHEAFFER 5-30 pen in jade green, gold filled trim. $60-75

1932 SHEAFFER long *Balance Lifetime*, bright jade green, gold filled trim, lever filler. $120-145

1931 SHEAFFER full size black & cream pearl pencil, gold filled trim, $75-85

1931 SHEAFFER, *Lifetime*, black, gold filled trim, lever filler. $45-60

1932 SHEAFFER, Black
Dot Midget pen in green
pearl marble, lever filler.
$55-70

1932 SHEAFFER, jade
green telephone dialer
pen, gold filled trim,
lever filler. $60-85

1932 SHEAFFER *Balance
Lifetime*, black, gold filled trim,
lever filler. $60-85

1932 SHEAFFER *Balance Lifetime*, black & cream pearl, gold filled trim, lever filler, barrel color is perfect. $125-150

1932 SHEAFFER full size *Balance Lifetime*, red veined grey marble Radite, gold filled trim, lever filler. $375-425

1932 SHEAFFER *Balance* 3-25, blue & black marble Radite, gold filled trim, lever filler. $200-275

1932 SHEAFFER *Lifetime*, black
with 14kt gold trim, lever filler.
$160-200

1933 SHEAFFER,
Lifetime, black, gold filled
trim, lever filler. $50-70

1933 SHEAFFER,
Lifetime, black, gold filled
trim, lever filler. $50-70

1932 SHEAFFER full
size *Lifetime*, black with
14kt gold trim, lever
filler. $175-300

1934 SHEAFFER golf pencil, gold filled trim, $20-30

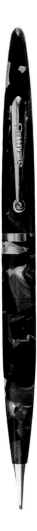

1934 SHEAFFER grey pearl pencil, nickel plated trim. $25-45

1934 SHEAFFER Balance *Lifetime*, black, gold filled trim, lever filler. $50-70

1934 SHEAFFER full size *Balance Lifetime*, black with 14kt gold trim, lever filler. $175-300

1935 SHEAFFER *Lifetime*, black with mother-of-pearl chips, gold filled trim, lever filler. $100-145

1936 SHEAFFER, #3, black with inlaid mother-of-pearl chips, gold filled trim, lever filler. $135-155

1936 SHEAFFER full size *Balance Lifetime*, red pearl Radite, gold filled trim, lever filler. $375-425

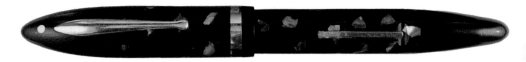

1936 SHEAFFER full size *Lifetime*, black with mother-of-pearl chips, gold filled trim, lever filler. $385-425

1936 SHEAFFER, *Lifetime* 1000, black, gold filled trim, name on barrel, lever filler. $50-60

1937 SHEAFFER, roseglow two tone, gold filled trim, piston filler. $45-55

1937 SHEAFFER, roseglow two tone, gold filled trim, lever filler. $125-175

1937 SHEAFFER, golden brown two tone, gold filled trim, lever filler. $60-80

1939 SHEAFFER 875 *Lifetime* pen, gold stripe, piston filler, gold filled trim. $25-30

1941 SHEAFFER 500 Vigilant pen in green pearl stripe, gold filled trim, lever filler. $50-65

1939 SHEAFFER full size *Lifetime*, black, gold filled trim, lever filler. $135-175

1939 SHEAFFER *Lifetime*, black, gold filled trim, unusual grooved band, lever filler. $60-75

1945 SHEAFFER Crest
Triumph *Lifetime* pen &
pencil, black with 14kt gold
caps, lever filler, $395-450

1945 SHEAFFER 1000
Tuckaway *Lifetime* pen & pencil,
black, piston filler, gold filled
trim. $80-100

1946 SHEAFFER
Tuckaway *Lifetime* pen,
maroon with chrome &
gold filled cap, plunger
with sac filler. $45-55

1948 SHEAFFER, Lifetime
Triumph, black, gold filled trim,
lever filler. $55-60

1948 SHEAFFER 875 pen in
brown, gold filled trim, lever filler.
$30-45

1946 SHEAFFER Triumph
pen, black, lever filler, gold
banded Lustraloy cap. $75-90

1949 SHEAFFER 1500 Triumph
pen, black, lever filler, gold filled
trim. $75-90

1950 SHEAFFER, *Lifetime*.
Triumph Sentinal, blue with chrome
& gold filled cap, plunger with sac
filler. 470-80

1952 SHEAFFER Clipper Snorkel,
chrome & gold filled cap, red Barrel.
$40-55

1949 SHEAFFER *Stratowriter*
Ballpoint, gold filled metal.
$35-45

1952 SHEAFFER, *Lifetime* Triumph
Snorkel pen & pencil in 14kt gold,
first year of Snorkel issue. $490-550

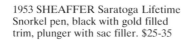
1953 SHEAFFER Saratoga Lifetime
Snorkel pen, black with gold filled
trim, plunger with sac filler. $25-35

1959 SHEAFFER PFM I,
Lifetime pen, black with
chrome plated trim, plunger
with sac filler. $80-110

1953 SHEAFFER Sentinal
Lifetime Snorkel pen, black
with chrome & gold filled
cap, plunger with sac filler.
$35-45

1959 SHEAFFER, *Lifetime PFM II*, blue, chrome cap, plunger with sac filler. $125-145

1960 SHEAFFER, Imperial pen, blue with brushed chrome cap, plunger with sac filler. $40-60

1959 SHEAFFER *PFM V*, Lifetime pen, black with gold filled cap, plunger with sac filler. $325-375

1959 SHEAFFER *PFM III*, blue with gold filled trim. $175-225

1907 H. B. SMITH pen, made in Janesville, Wisconsin. Harry Smith, a former Parker employee started his own pen company and did quite well. Eyedropper filler. $105-120

1910 STERLING middle joint pen, red & black hard rubber. Sterling was the leader in middle joint pens. Quality was exceptional. $185-225

1910 STERLING middle joint pen, larger #5 pen, black hard rubber with a silver snake clip. $195-245

1910 STERLING pen, standard eyedropper style, black hard rubber. $60-75

1916 SWAN #4 S.F. pen, Mabie Todd & Co., black, gold filled trim, lever filler (rare split two-part lever only made for a year). $135-160

1915 MABIE TODD & CO. pen, marked "The Roose", black chased hard rubber, coin filler. $95-145

1915 SWAN SF1, 1000/60, black chased hard rubber, lever filler. $50-65

1917 SWAN #2 S.F. pen, Mabie Todd & Co., black, nickel plated trim, lever filler. $50-65

1919 SWAN pen, M. T. & Co., sterling silver hand-engraved, lever filler. $200-245

1925 SWAN pen, M. T. & Co., sterling silver, lever filler. $195-225

1921 SWAN #71 pen, Mabie Todd & Co., black, gold filled trim, lever filler. $35-50

1920 MABIE TODD & CO. pen, stamped as if the maker was "The Burrows Brothers Company, Cleveland", black chased hard rubber, lever filler. $50-65

1925 SWAN Pen, M. T. & Co., 14kt hand-engraved gold, lever filler. $450-600

1926 SWAN sterling silver engraved design pen & pencil, Mabie Todd & Co., lever filler. $525-575

1926 SWAN 44 Eternal, red & black woodgrain, hard rubber, Mabie Todd & Co., lever filler, gold filled trim. $250-300

1926 MABIE TODD & CO. Blackbird pen, black chased hard rubber, gold filled trim, lever filler. $65-80

1926 SWAN pencil, Mabie
Todd & Co., perfect jade
green, gold filled trim. $50-65

1927 SWAN 54 ETN pen,
Mabie Todd & Co., lapis
blue with gold filled trim,
lever filler. $145-190

1926 SWAN full size pencil,
Mabie Todd & Co., brilliant
lapis blue with gold filled
trim. $125-165

1927 SWAN pen, Mabie Todd & Co., rare color bronze veined creamy silver pearl "Lizard", gold filled trim, lever filler, made in England. $200-225

1927 SWAN pen, Mabie Todd & Co., rare color bright orange marble with black highlights and gold filled trim, lever filler. $300-350

1927 SWAN #242/55 pen, Mabie Todd & Co., black veined light blue pearl, gold filled trim, lever filler, made in England. $220-245

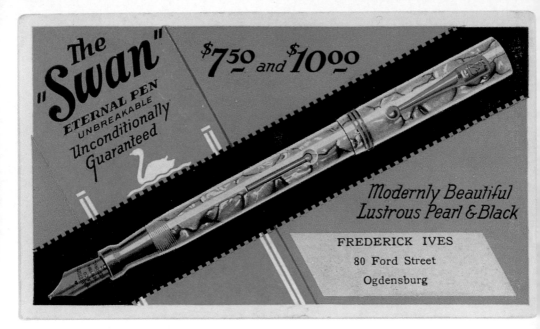

1929 SWAN Blotter, Mabie Todd & Co., five colors, great graphics. $35-40

1931 SWAN Eternal pen/pencil combo, black with green pearl marble inserts, gold filled trim, lever filler. $300-385

1932 SWAN "Leverless" pen & pencil, black chased hard rubber, Mabie Todd & Co., gold filled trim. $175-200

1933 SWAN pen, Mabie Todd & Co., red veined black, gold filled trim, lever filler. $175-225

1935 SWAN 1060 "Leverless", black, gold filled trim. $90-115

1933 SWAN desk pens (2) black, Mabie Todd & Co., lever filler, gold filled trim. $70-90

1937 SWAN "Leverless" rare grey-green lizard twist filler pen, gold filled trim. $295-340

1938 SWAN desk pen by Mabie Todd & Co., black striped gold pearl, gold filled trim, lever filler. $75-90

1939 SWAN pencil, M. T. & Co., 14kt pink gold. This is a rare design. $350-400

1949 SWAN #1 "Leverless", black, gold filled trim. $55-70

1949 MABIE TODD & CO. Blackbird demonstrator pen, gold filled trim, "Leverless". $135-165

1926 LOUIS THOMAS & SONS, 14kt heavy
gold, lined pattern, lever filler. $300-375

1910 Unknown pen, black chased hard
rubber, eyedropper filler, nice quality. $20-30

1950 SWAN 3230, grey, gold
filled trim, Italic point, lever
filler. $40-50

1918 Unknown European safety
pen, black chased hard rubber,
copy of Waterman #45. $85-100

1929 Unknown large pen, nice cream & brown woodgrain color, lever filler, gold filled trim. $40-50

1936 Unknown pen, sterling silver with raised ring design, European type twist filler. $100-135

1965 Unknown 18kt gold ballpoint pen (no cartridge), made in Italy, very high quality & an ounce-plus of 18kt gold. $200-225

Unknown English bulb filler, gold & black pearl marble, high quality, gold filled trim. $90-110

1948 UNIC, black, gold filled trim, made in France, Vacumatic filler. $75-90

1946 VENUS pen in black striped red & grey pearl, gold filled trim, lever filler. $25-30

1923 WAHL gold filled pen, Greek key design, lever filler. $65-80

1924 EVERSHARP hand-engraved sterling silver pencil. $110-140

1924 WAHL #4 pen in sterling silver, lever filler, rare pattern. $200-250

1925 WAHL #4 pen, gold filled, unusual wavy lined pattern, lever filler. $130-155

119

1925 WAHL #2 lady's pen, gold filled, chevron pattern, lever filler. $45-65

1925 WAHL #3 pen, gold filled, unusual wavy lined pattern, lever filler. $75-90

1925 WAHL #2 in red & black woodgrain, gold filled trim, lever filler. $40-50

1925 WAHL miniature in gold filled metal, lever filler. $60-80

1925 WAHL #2 pen in gold filled metal, lever filler. $50-60

1925 WAHL pen, red & black rosewood pattern. $55-65

1925 EVERSHARP LARGE #75 pencil in gold filled metal. $100-125

1927 WAHL full sized black chased hard rubber "Signature" pen with roller ball clip, gold filled trim, lever filler. $295-340

1927 EVERSHARP pencil, black chased hard rubber. $60-80

1927 WAHL #4 pen in orange chased hard rubber, gold filled trim, lever filler. $250-300

1927 EVERSHARP roller ball pencil in pearl & black. $50-60

1929 WAHL, red & black woodgrain, with roller ball clip, gold filled trim, lever filler. $135-155

1929 WAHL #0 Lady's pen, gold filled, Greek key pattern, lever filler. $65-80

1930 WAHL-EVERSHARP full size Gold Seal *Personal* Point deco band roller ball clip pen in jade green, lever filler, Gold Seal point. $400-575

1930 WAHL-EVERSHARP pen in red & black woodgrain, gold filled trim, lever trim. $45-60

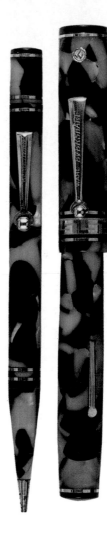

1930 EVERSHARP pencil in
bronze & green marble, gold
filled trim. $30-40

1930 WAHL-
EVERSHARP oversize
Gold Seal *Personal Point*,
deco band pen & pencil,
cream & black, gold filled
trim. $460-600

1930 WAHL desk pen in salmon
marble, lever filler. $55-75

1931 WAHL-EVERSHARP Gold Seal *Personal* Point *Equipoise*, black, lever filler, gold filled trim. $250-300

1931 WAHL-EVERSHARP oversize Gold Seal *Personal Point*, deco band pen, black, gold filled trim. $350-450

1932 WAHL-EVERSHARP large Gold Seal Convertible pen, black, gold filled trim. This model, with the addition of the pencil end, became a pen/pencil combination, and with the taper end it became a desk pen. $350-400

1933 WAHL-EVERSHARP "Midget" pen & pencil, red burgundy marble, gold filled trim. $45-65

1935 WAHL-EVERSHARP *Doric*, garnet red pearl, lever filler, gold filled trim. $250-285

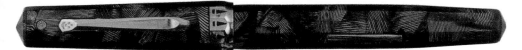

1935 WAHL-EVERSHARP *Doric*, oversize, red pearl plastic, lever filler. $375-475

1936 WAHL-EVERSHARP *Doric*, gold pearl plastic, lever filler. $295-340

1936 WAHL-EVERSHARP *Doric*, black plastic, lever filler. $155-200

1936 WAHL-EVERSHARP *Doric* pen, green veined brown & gold pearl, gold filled trim. $165-200

1937 WAHL-EVERSHARP *Doric*, burgundy lined pearl, safety shut off, gold filled trim. $135-155

1938 WAHL-OXFORD in gold pearl, gold filled trim, end bulb filler. $40-50

1938 EVERSHARP *Coronet* style pencil, green pearl, nickel plated trim. $35-45

1938 EVERSHARP *Coronet* style pencil, gold pearl, nickel plated trim. $35-45

1938 WAHL-EVERSHARP "Premium Model" pen, black & grey pearl, nickel plated trim. This model was made for company premiums and is marked "Specify Mead's". A very Un-Wahl color. $125-175

1939 WAHL-EVERSHARP Varsity pen, brown/black striped pearl, lever filler, gold filled trim. $85-100

1942 EVERSHARP "Gold Award" *Skyline* pen, blue with golf filled metal, line pattern, and rare two tone nib. A variation in the Gold Award design. $140-165

1942 EVERSHARP "Gold Award" *Skyline* pen & pencil, gold filled metal, chevron and line pattern. This is a rare Skyline pen & pencil. $300-350

1942 EVERSHARP *Skyline* pen, burgundy with gold filled derby & wide gold filled band. These wide band pens with gold filled derbies were the deluxe models. Lever filler, gold filler trim. $85-100

1942 EVERSHARP *Skyline* pen, dark green barrel, gold filled cap. $45-65

127

1943 EVERSHARP *Fifth Avenue*, maroon with gold filled trim, lever filler. $35-45

1943 EVERSHARP *Fifth Avenue*, dark grey with 14kt gold cap, lever filler. $175-200

1944 EVERSHARP "Executive" *Skyline* pen, oversized pen green with gold stripe cap and rare two tone nib. $200-225

1943 EVERSHARP *Fifth Avenue*, lady's model, black with 14kt gold cap, lever filler. $175-200

1944 EVERSHARP *Skyline* pen, fully bronze stripe & unusual "WAHL" clip. $95-120

1946 EVERSHARP *Skyline* pen, blue with gold filled cap & derby, lever filler, gold filled trim. $75-90

1946 EVERSHARP *Skyline* pen, gold filled cap, lever filler, gold filled trim. $75-90

1946 EVERSHARP *Skyline*, grey with gold filled trim, lever filler. $45-55

1946 EVERSHARP *Skyline*, maroon with gold filled cap, lever filler. $65-85

1946 EVERSHARP *Skyline*, brown with gold filled cap, lever filler. $65-85

1946 EVERSHARP *Skyline* pen & pencil, black with gold filled & derby, lever filler, gold filled trim. $75-90

1946 EVERSHARP *Skyline* pen, stainless steel cap on black barrel, lever filler. $400-450

1946 EVERSHARP *Skyline* pen & pencil, gold filled caps, green barrel, lever filler. $100-120

1947 EVERSHARP "$64"
Fifth Avenue pen & pencil,
black with 14kt gold caps,
lever filler. $225-275

1950 EVERSHARP
"Loewy" pen, designed by
Raymond Loewy, maroon
with chrome & gold filled
cap, lever filler. $50-60

1950 EVERSHARP
"Loewy" pen, designed by
Raymond Loewy, black
with chrome & gold filled
trim cap, lever filler. The
Loewy pens are 1950s
"retro" design (sort of a
1950s Art Deco style). $50-
60

1948 EVERSHARP Envoy
in gold filled metal, lever
filler. $25-35

131

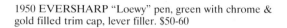

1950 EVERSHARP "Loewy" pen, green with chrome &
gold filled trim cap, lever filler. $50-60

1950 EVERSHARP "Loewy" pen, blue
with chrome & gold filled trim wide band
cap, lever filler. $60-70

1952 EVERSHARP "Symphony" pen & pencil, designed by Raymond Loewy, black with
chrome & gold filled trim cap, lever filler. $65-80

1952 EVERSHARP slipper-shape cap
"Loewy" pen, maroon with chrome & gold
filled cap, lever filler. $45-55

1952 EVERSHARP "Symphony" pen & pencil, designed by Raymond Loewy, black with brushed steel & chrome cap, lever filler. $50-65

1935 WALTHAM "Indian" pen/pencil combo, blue, white, green & maroon, gold filled trim & nickel plated trim, lever filler. $30-35

1898 WATERMAN #12, red & black wood grain, eyedropper filler. This pen has the early two line imprint without the 1903 globe and the early three fissure feed. $100-150

1903 WATERMAN #22 pen, eyedropper filler, comes with eyedropper, instructions and carrying case (case is in rough condition). $200-250

1905 WATERMAN #12, red & black
wood grain, eyedropper filler. This pen
has the imprint with the 1903 globe and
the 1899 spoon feed. $100-125

1905 WATERMAN #12,
999 FINE silver overlay pen,
eyedropper filler. $600-750

1908 WATERMAN #15 pen, black
chased hard rubber, nickel plated
trim, eyedropper filler. $125-175

1910 WATERMAN #15 POC,
black eyedropper filler with a
hand-engraved 14kt gold band.
This was the first screw thread
model. $185-225

1910 WATERMAN #512 1/2 VS 14kt gold Chatelaine smooth finish safety pen, fully gold-covered (no black hard rubber shows anywhere). $600-700

1910 WATERMAN hard rubber token in natural rubber color with wooden insert. $190-220

1910 WATERMAN #14 pen, black chased hard rubber, eyedropper filler. $50-60

1912 WATERMAN #0742 1/2V pen, black chased hard rubber, gold filled trim, "07" means gold filled band. $100-150

1912 WATERMAN #415 SF, SLEEVE FILLER pen with sterling silver filigree overlay. Large silver sleeve fillers (made from 1910-1915) are scarce. $1650-1900

1912 WATERMAN #12 SF, Sleeve Filler (made between 1910-1915), nickel plated trim. $200-250

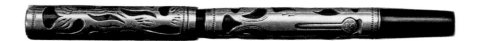

1914 WATERMAN #412 1/2 PSF, sterling silver overlay, lever filler. An early lever filler. $320-400

1915 WATERMAN #12 1/2 PSF, black with 14kt gold clip, lever & band, lever filler. The first lever filling models were designated "PSF" for Pocket Self Filler. $295-335

1915 WATERMAN #12 1/2 PSF pen, black hard rubber, gold filled trim, lever filler. $65-80

1917 WATERMAN #452 1/2, sterling silver "Oriental" pattern overlay, lever filler. The Oriental pattern, in silver, is one of the most lovely hand-engraved patterns. $1500-1800

1917 WATERMAN #0552 1/2, gold filled "Oriental" pattern overlay, lever filler. $850-950

1918 WATERMAN #42 safety pen, black chased hard rubber, eyedropper filler. $80-110

1918 WATERMAN #16 pen, black hard rubber, eyedropper filler, later "keyhole" nib. $150-175

1918 WATERMAN #42 1/2V safety pen, black chased hard rubber, eyedropper filler. $90-115

1920 WATERMAN #52X in black, nickel plated trim, lever filler. The 52X model is the size of a #55 pen with a #2 nib. It is rare. $215-245

1920 WATERMAN #412 1/2, eyedropper filler, ultra thin silver filigree pen. $385-435

1920 WATERMAN #52 1/2V pen, orange hard rubber, lever filler, gold filled trim. $95-120

1920 WATERMAN #442 1/2V, safety pen, eyedropper filler, barleycorn sterling silver pen, a very rare pattern. $400-475

1921 WATERMAN #0752 pen, black chased hard rubber, gold filled trim. Unusual to see the "0752" stamped on the end. $85-110

1922 WATERMAN #452 1/2 LEC sterling silver fully hand-engraved. $450-490

1922 WATERMAN black hard rubber pencil. $25-35

1922 WATERMAN #0512 1/2, gold filled "Pansy Panel" overlay, lever filler. $420-450

139

1922 WATERMAN
#552 1/2 LEC 14 kt
gold Sheraton pattern,
lever filler. $700-800

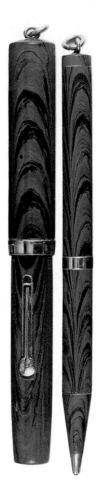

1923 WATERMAN #52V pen
& pencil, red & black ripple,
gold filled trim, lever filler.
$125-150

1922 WATERMAN
sterling silver engraved
filigree pencil. $75-85

1922 WATERMAN sterling silver
fully hand-engraved (crown &
clip) pencil. $85-100

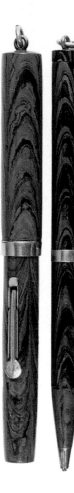

1924 WATERMAN 52 1/2V red & black ripple pen & pencil. $85-115

1924 WATERMAN #442 1/2V, safety pen, eyedropper filler, hand-engraved sterling silver. $250-300

1924 WATERMAN #452 1/2V, lever filler, hand-engraved silver pen. Waterman produced many of these beauties and every one is slightly different from the other. This is the higher quality model with the engraved lever. $100-160

1924 WATERMAN 55 red & black ripple, lever filler, gold filled trim. $200-265

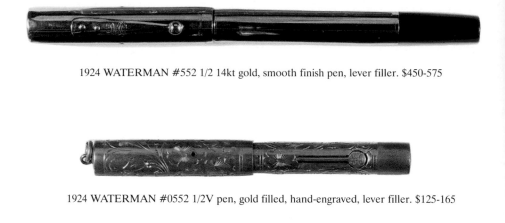

1924 WATERMAN #552 1/2 14kt gold, smooth finish pen, lever filler. $450-575

1924 WATERMAN #0552 1/2V pen, gold filled, hand-engraved, lever filler. $125-165

1924 WATERMAN #552 1/2 14 kt gold, smooth finish pen, lever filler, no engraving. $450-575

1924 WATERMAN #552 1/2 14kt gold, smooth finish overlay pen barrel, lever filler, no engraving, but the cap should also be gold covered or have a gold clip. $140-160

1924 WATERMAN #554 14kt gold, filigree pen, lever filler. $850-1000

1924 WATERMAN #452 sterling silver, filigree pen, lever filler. $350-450

1924 WATERMAN #452 1/2V pen, silver, gothic design, lever filler. $100-145

1924 WATERMAN #0552 1/2V, gold filled, gothic design, lever filler. $95-110

1925 WATERMAN #452 1/2V pen, silver, hand-engraved, lever filler. $100-145

1925 WATERMAN #52V pen, black chased hard rubber, nickel plated trim, lever filler. $35-40

1925 WATERMAN #452 1/2V
pen, sterling silver, lever filler.
$100-145

1925 WATERMAN
#0512 1/2V, gold filled
Sheraton design, lever
filler, in the rare clip
model. $145-175

1925 WATERMAN
#52, red & black
wood grain, no
band model, lever
filler, gold filled
trim. $100-125

1925 WATERMAN #52
pen, red & black ripple,
gold filled trim, lever
filler. $100-125

1925 WATERMAN #52 1/2V red
& black ripple, gold filled trim,
lever filler. $65-75

1925 WATERMAN #55 black
chased hard rubber pen, nickel
plated trim, lever filler. $100-130

1925 WATERMAN
#56 red & black
ripple, gold filled
trim, lever filler.
$300-350

1926 WATERMAN #452 1/2V, vest pocket pen, lever filler, sterling silver, an unusual pen in the clip model. $240-285

1926 WATERMAN #452 1/2V pen, silver, lever filler. $65-80

1926 WATERMAN 554 Sheraton pen & large pencil, 14kt gold, lever filler. $1000-1200

1927 WATERMAN #5 pen, red & black ripple, gold filled trim, lever filler. $150-175

1928 WATERMAN #01955 red & black ripple, gold filled trim, lever filler. $235-265

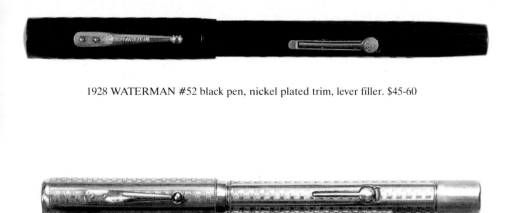

1928 WATERMAN #52 black pen, nickel plated trim, lever filler. $45-60

1928 WATERMAN #554 LEC gothic design, 14kt, lever filler, "LEC" stands for Lower End Covered. $800-1000

1928 WATERMAN #94 pen, olive ripple, lever filler. $275-325

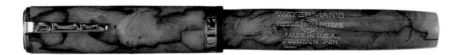

1930 WATERMAN #94 red/grey marble, lever filler, "Purple" point, nickel plated trim. $110-150

1930 WATERMAN *Lady Patrica*, red veined cream (agate), gold filled trim, lever filler. $80-110

1932 WATERMAN
(purple dot) #7 black pen,
"Purple" point, gold filled
trim, lever filler. $150-175

1935 WATERMAN #3,
red spotted black & grey
pearl, nickel plated trim,
lever filler. $45-60

1933 WATERMAN pen/
pencil combination, lever
filler. This is one of the
very few pen/pencil
combinations that
Waterman made. $625-700

1935 WATERMAN #3V, brown lined grey pearl,
nickel plated trim, lever filler. $55-65

1935 WATERMAN business card for "J. C. Baker" at the 140 Thomas
Street, Newark address, a rare piece of ephemera. $35-45

1936 WATERMAN #403, sterling silver overlay
pen & pencil, lever filler. $850-1000

1947 WATERMAN pen & pencil, burgundy woodtone color, gold filled trim, French made, lever filler. $175-210

1939 WATERMAN #513J golden pearl, lever filler, gold filled trim. $150-180

1939 WATERMAN light green pearl
desk pen, lever filler. $95-115

1940 WATERMAN Hundred Year pencil, 5 inches long,
translucent green, gold filled trim. $105-130

1940 WATERMAN Thoroughbred pencil, blue veined grey pearl, gold filled trim. $45-50

1941 WATERMAN golden pearl pen & pencil, lever filler. $120-145

1942 WATERMAN golden cream pearl, gold filled trim, lever filler. $65-80

1943 WATERMAN *Emblem* pen in black, gold filled trim, lever filler, slip cap style. $50-60

1943 WATERMAN Commando black pen, gold filled trim, lever filler. $55-65

1948 WATERMAN Stateleigh *Taperite*, black with gold filled cap, lever filler, gold filled trim. $45-50

1956 WATERMAN "CF" pen and pencil, black with gold filled caps. $75-90

1958 WATERMAN black with chrome & gold filled cap, unusual Aerometric filler. $50-60

1965 WATERMAN DG pen, silver with black, cartridge filler. $70-80

1965 WATERMAN pen, black, gold filled trim, cartridge filler. $60-70

L.E. WATERMAN #404, ca. 1895-1910. Sterling silver "Repousse" design overlay eyedropper filled pen. $14,000-16,000

L.E. WATERMAN, ca. 1905-1915. The "Check-book" pen with a sterling silver overlay. Made to accompany a notepad or checkbook, eyedropper filled. $2500-3500

L.E. WATERMAN, ca. 1936-1939. The "Bayleaf" pen with a sterling silver overlay. This is also called the "Rabbit Ear" design. $1500-2000

1929 WAVERLEY "Cameron" pen, black, eyedropper filler, gold filled trim, 14kt gold point, made in England. $25-30

1920 YANKEE pen made by PenCraft of Chicago, black chased hard rubber, lockdown lever filler. $30-40

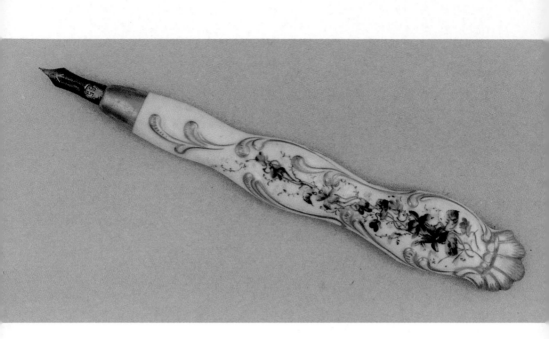

1800s Unknown French porcelain hand-painted & gilt dip pen, with floral motif. $85-145

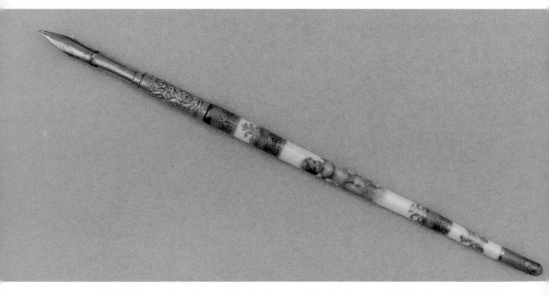

1800s Unknown French porcelain hand-painted & gilt dip pen with cherub motif. $250-300

154

1850 Gorham & Co. sterling silver retractable propelling pencil with riding motif.
$275-325

1860 Unknown bone-handled propelling pencil with Stanhope of the Palais du
Trocadero. $35-45

1860 Sterling silver propelling pencil. $75-100

1860 W. S. Hicks red hard rubber combination propelling pencil & dip pen. Pen is labeled with
Goodyear 1855 patent and fitted with Aleeri Bagley nib. $125-175

1870 Mabie, Todd & Co. black chased hard rubber propelling pencil. $35-50

155

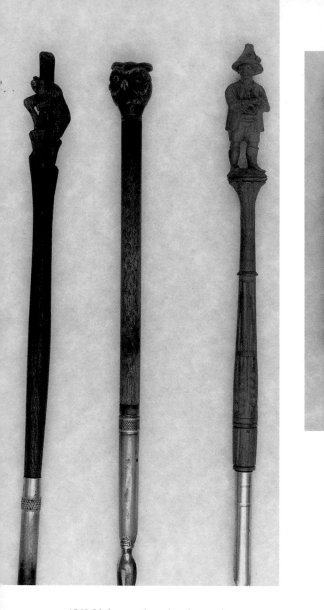

1870 Unknown ivory and 14kt gold inlaid propelling pencil. $200-225

1860 Unknown three hand-carved wooden figural dip pens, probably German or Austrian. $50-80

1870 L. W. Fairchild hand-engraved 14kt gold combination dip pen and retractable pencil. $350-400

1870 E. S. Johnson black hard rubber and 14kt gold floral inlay propelling pencil. $90-140

1870 Mabie, Todd & Co. wood and gold filled engraved handled retractable dip pen. $75-100

1875 Unknown unusual suede covered dip pen in the form of a cigar. $95-140

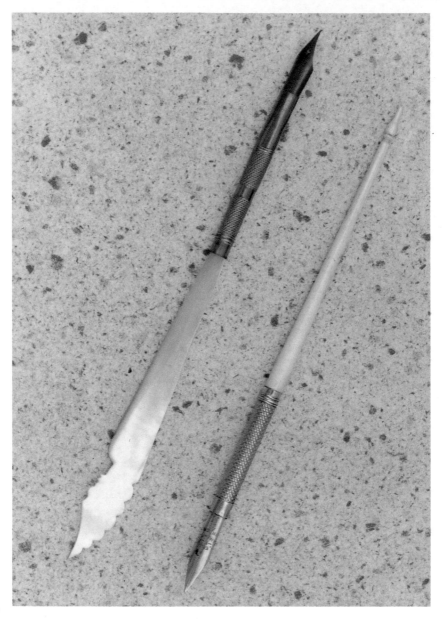

1880 Unknown ivory and mother-of-pearl handled dip pens. $40-75

1880 John Holland red & black mottled hard rubber dip pen. Pen is shown in open position. $100-150

1900 Unger Bros. sterling silver pen stand decorated with cherubs. $400-500

1880 Aikin Lamber alternating abalone & mother-of-pearl retractable propelling pencil. $45-80

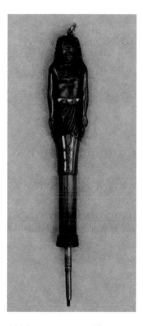

1920 Unknown advertising pencil for Heidseick. Monopole Champaigne. $75-110

1925 copper propelling pencil with enameled Egyptian motif, ca. 1925. $85-150

1880 Unknown ivory handled dip pen with Edward Todd nib. $75-125

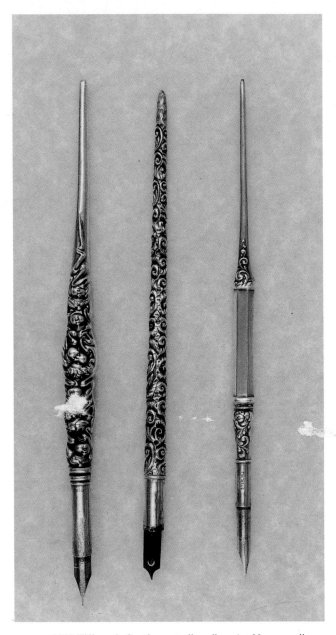

1900 Tiffany & Co. three sterling silver Art Nouveau dip pens. $200-300

1930 Eagle Home Run King wooden handled dip pen in the shape of a baseball bat. $40-65